IMAGES
of America

FIREFIGHTING IN
CHARLOTTE

IMAGES
of America

FIREFIGHTING IN
CHARLOTTE

Shawn Royall

ARCADIA
PUBLISHING

Published by Arcadia Publishing
Charleston SC, Chicago IL, Portsmouth NH, San Francisco CA

Library of Congress Catalog Card Number: 2007922261

For all general information contact Arcadia Publishing at: Telephone
843-853-2070
Fax 843-853-0044
E-mail sales@arcadiapublishing.com
For customer service and orders:
Toll-Free 1-888-313-2665

Visit us on the Internet at www.arcadiapublishing.com

I would like to dedicate this book to all firefighters—past, present, and future—for carrying on the rich traditions, brotherhood, pride, duty, loyalty, and honor that this job demands every day. To my wife, Gretchen, and our daughter Addison, thank you for your devotion and continuous support. I will love you, always and forever.

CONTENTS

Acknowledgments 6

Introduction 7

1. 1880–1920 9

2. 1920–1940 27

3. 1940–1960 43

4. 1960–1980 69

5. 1980–1990 101

Further Information 126

ACKNOWLEDGMENTS

I would like to extend my appreciation to the following for their invaluable contribution to this project: the Public Library of Charlotte and Mecklenburg County, Robinson-Spangler Carolina Room; the *Charlotte Observer* newspaper, Archives Department (unless otherwise noted, images were obtained from the *Charlotte Observer* archives); Charlotte Fire Department History Book, 2000 edition; Battalion Chief Harley Cook (CFD retired); Capt. Rob Brisley, CFD public information officer; fire investigator Paul Wilkinson, CFD Arson Investigation Task Force; engineer Glen Garris, CFD Ladder Company No. 1; engineer John Foster (CFD retired), president of Charlotte Firefighters Local 660; UNC-Charlotte, J. Murrey Atkins Library, Special Collections; the Charlotte Museum of History; and the Charlotte-Mecklenburg Fire Museum and Education Center.

INTRODUCTION

Throughout the history of the Charlotte Fire Department, its story has mirrored that of the city it serves—one of constant change. Charlotte has grown from a trading crossroads on the site of the square at Trade Street and Tryon Street into one of the largest cities in the United States. Along the way, citizens have maintained fire protection since the earliest days of leather buckets and hand-drawn fire engines. The discovery of gold, railroads, cotton, textile mills, and the banking industry have all shaped the city of Charlotte into the metropolis it is today. Fire is an unforgiving enemy, thus the fire department proudly continues today, stronger and more determined than ever to protect the citizens of Charlotte.

Firefighting in Charlotte is a brief photographic glimpse into the past and attempts to give the reader an understanding of how the Charlotte Fire Department has evolved. In the years leading up to 1880, Charlotte was growing vigorously and on the verge of an industrial and commercial boom. The town's strategic railroad location brought together local businessmen and entrepreneurs, and created a cultural melting pot. Charlotte, with just over 7,000 inhabitants, was ruled by the cotton mills of the textile industry. According to history, a group of leading Scotch-Irish citizens voluntarily declared their defiance of Britain's royal authority by drafting and signing the Mecklenburg Declaration of Independence on May 20, 1775, more than a year before the official declaration was signed on July 4, 1776. Over 100 years later, the future generations of those patriotic citizens of Charlotte and Mecklenburg County were protecting themselves from the dangers of fire with volunteer fire companies.

Before 1887, fire protection was accomplished by several volunteer fire companies that were composed of both black and white citizens. Fire companies such as the Hornets, the Pioneers, and the Neptunes have all thundered down the streets of Charlotte at the shout of "fire" or the toll of the fire bell at the square. In the department's short, 120-year existence, from 1887 to 2007, the Charlotte Fire Department has grown from volunteer firemen, two firehouses, horse-drawn steam fire engines, and water drawn from cisterns to 38 firehouses, 60 fire trucks, and over 1,000 male and female firefighters. For such a long time, little was saved to give much insight into the daily operations of the CFD, but with these photographs, hopefully time can stand still for a brief moment. Although great attempts have been made to verify dates, locations, names, and events when rediscovering history, mistakes can occur.

One

1880–1920

In the days preceding the Civil War until 1887, there were several well-established volunteer fire companies protecting Charlotte, including the Hornet Fire Engine Company, the Pioneer Fire Engine Company, and the Independent Hook and Ladder Company. The Neptune Hand Engine Company was composed mostly of former or current slaves, and other short-lived African American companies came and went, such as the Yellow Jacket Fire Company and the Dread Naught Fire Company. At the ring of the fire bell, citizens would throw leather buckets into the streets and the volunteers would grab them while running to the fire with their hand-drawn fire engine. Up until 1882, when a waterworks was installed, the volunteers relied on underground cisterns that collected rainwater to draw water from with suction hose. Eventually hand-drawn pumpers gave way to horse-drawn steam engines, and the volunteers struggled to keep up with the ever-expanding town.

Early in the summer of 1887, the board of alderman voted to hire a full-time fire marshal. The volunteers did not disagree with the decision per se, but asked the alderman to consider their chief for the job. In July, when the alderman refused and hired a less experienced man, the volunteers turned over all their equipment to the city and three of the companies disbanded, leaving only the Neptunes to temporarily protect the city. On August 1, 1887, the Charlotte Fire Department was officially organized by the board of aldermen.

By 1900, electric streetlights replaced gaslights and electric streetcars were carrying citizens through the town of nearly 18,000 people. Fire protection consisted of an established waterworks with 147 hydrants, 26 pull boxes, 12 paid firemen, 40 paid on call, and 9 horses. Equipment included one Clapp and Jones double-piston steam fire engine, one Paterson, New Jersey–made steam fire engine, one extension hook-and-ladder truck, two hose wagons, and one hose reel. Cotton, rubber-lined fire hose replaced brass-riveted leather hose, and fire duty remained consistent.

In 1911, the department consisted of 3 firehouses, 17 men, 13 horses, a Gamewell alarm system with 48 non-interfering boxes, and an automatic fire bell in a tower beside headquarters. In 1917, the department purchased its first motorized apparatus, the last of the fire horses were retired from duty, a two-platoon work schedule was developed, and the city of Charlotte prepared for the Roaring Twenties.

No 1 Pioneer Company

The Hornet Steam Engine Company No. 1, organized in 1867, is pictured in front of their quarters located at 222 East Trade Street in the 1880s. To the left is their horse-drawn Clapp and Jones steam fire engine and to the right is the hose wagon. The company mascot stands proudly in front alongside the firefighters. (CFD Fire Museum.)

OFFICERS.

W. H. TREZEVANT, President.

W. R. MYERS, Jr.....1st Director.   M. S. FRAZIER,.........Chief Engineer
R. E. MILLER.........2nd   "   W. B. TAYLOR,.........1st Asst.   "
J. W. WILSON.........3rd   "   JNO. F. FAIRINGTON..2nd   "
R. M. CRAWFORD, 4th   "   GEO. RIGLER...........Sexton.
ROBT. R. RAY,......Recording Secretary. A. D. COWLES........Financial Secretary.
                    D. C. MALLOY, Treasurer.

ROLL.

Austin, P. W.
Barnhardt, J. R.
Clarkson, T. S.
Cochrane, W. R.
Davidson, E. L.
Eagle, P. C.
Gibson, G. I.
Graham, W. C.
Griffith, F.
Henney, J. P. J.
Jones, Sol. B.
Johnston, W. R.
Kendrick, J. M.
Kerstek, F. W. T.
Liddell, J. W.
Little, T. H.
McGill, J. W.
McLaughlin, Jas.
McKay, Paul.

ROLL.

McCrasen, J. W.
McCorns, J. W.
Miller, Chas.
Overman, C. H.
Phifer, Geo. M.
Quinn, M. C.
Ray, Ed. C.
Ross, R. C.
Rigler, D. M.
Stephens, J. W.
Stokes, R. F.
Springs, A. B.
Spratt, C. E.
Selby, J. E.
Vance, C. N.
Wilson, L. H.
Wilson, S. W.
Wilkinson, W. T.
L. R. Wriston.

Dr. T. J. Moore, Surgeon.

Rev. E. H. Harding, Chaplain.

HORNET STEAM FIRE Co. No. 1,

CHARLOTTE, N. C.,

To Independent Columbia SC

MAY 20, 1875.

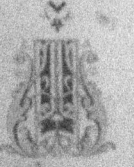

The certificate seen in this photograph lists the officers and members of the Hornet Steam Engine Company No. 1 in 1875. All dressed in their parade-best uniform, the officers, with their trumpets, are ready to call out orders, and the firemen are ready with their Clapp and Jones steam engine and hose reel. (CFD Fire Museum.)

This monument, erected in 1883, is located in the Elmwood Cemetery near present-day uptown Charlotte. It is a lasting tribute put up by the volunteer firemen of Charlotte honoring their dead comrades. Many of the prominent members of society were a part of the volunteer fire companies, and money was reportedly raised by the fire department ladies' auxiliary to build the statue and have it prominently displayed. (Royall.)

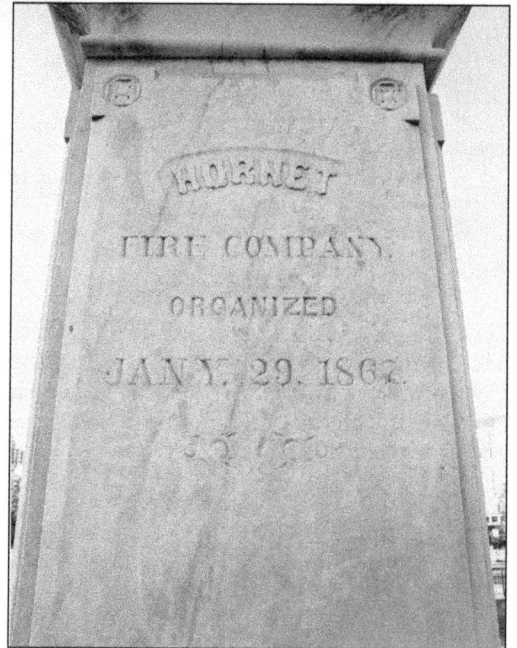

Above left, this close-up view of the firemen's monument in Elmwood Cemetery records the date it was erected along with the year that the fire department was organized. Although it has been determined that a fire department was operating in Charlotte well before the 1850s, an organized, paid department was not established until 1887. Above right, another view taken of the firemen's monument in Elmwood Cemetery honors the Hornet Steam Fire Engine Company, which was organized in 1867 and served until 1887. (Royall.)

12

The Pioneer Fire Engine Company, organized in 1874, poses with their hand-drawn hose reel in front of quarters in 1883. The men holding trumpets in this photograph were likely the company officers. They would shout commands, using these trumpets to carry their voice over the noise at a fire scene. (CFD Fire Museum.)

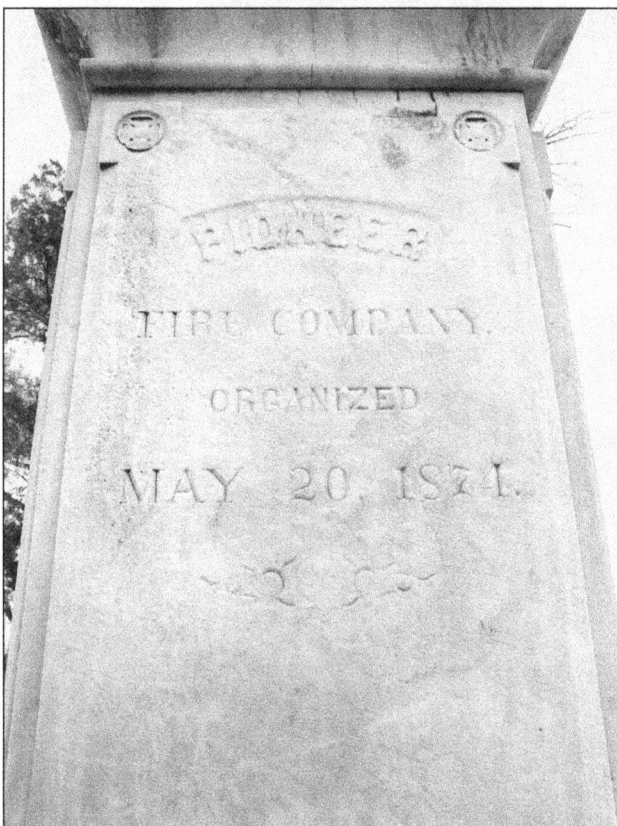

Pictured is the Pioneer Fire Engine Company turning out for a run in the mid-1880s. The harness for the horses to pull the steamer or hose wagon were suspended from above on a pulley system that would lower onto the horse once he was moved into position. The fireman would make the connection, and the company was ready. The floors of the firehouse were often made with brick or cobblestone and could become slick when wet. When laying the stone for these floors, grooves were created and placed horizontally to the apparatus door in order for the powerful horses to gain traction when pulling the heavy apparatus loaded with firemen. (Garris.)

This view of the firemen's monument in Elmwood Cemetery shows the side that honors the Pioneer Fire Company, organized in 1874. (Royall.)

The Neptune Fire Company, organized in 1864, was composed solely of slaves. Pictured here are two firemen with their chemical hose wagon in front of quarters. The Neptunes were a strongly competitive group and, although often excluded from joining the ranks of the other companies, were quite adept at holding their own at fires as well as at firemen competitions held at the annual Mec Dec Day on May 20, which could see as many as 35 competing fire companies. During July 1887, when the other volunteer companies disbanded and turned their equipment over to the city, the Neptunes continued providing fire protection. They would be the only company for nearly a month until the board of aldermen voted to create a paid fire department. The Neptunes formally disbanded in 1905, thus ending the last of the volunteer fire companies. (CFD Fire Museum.)

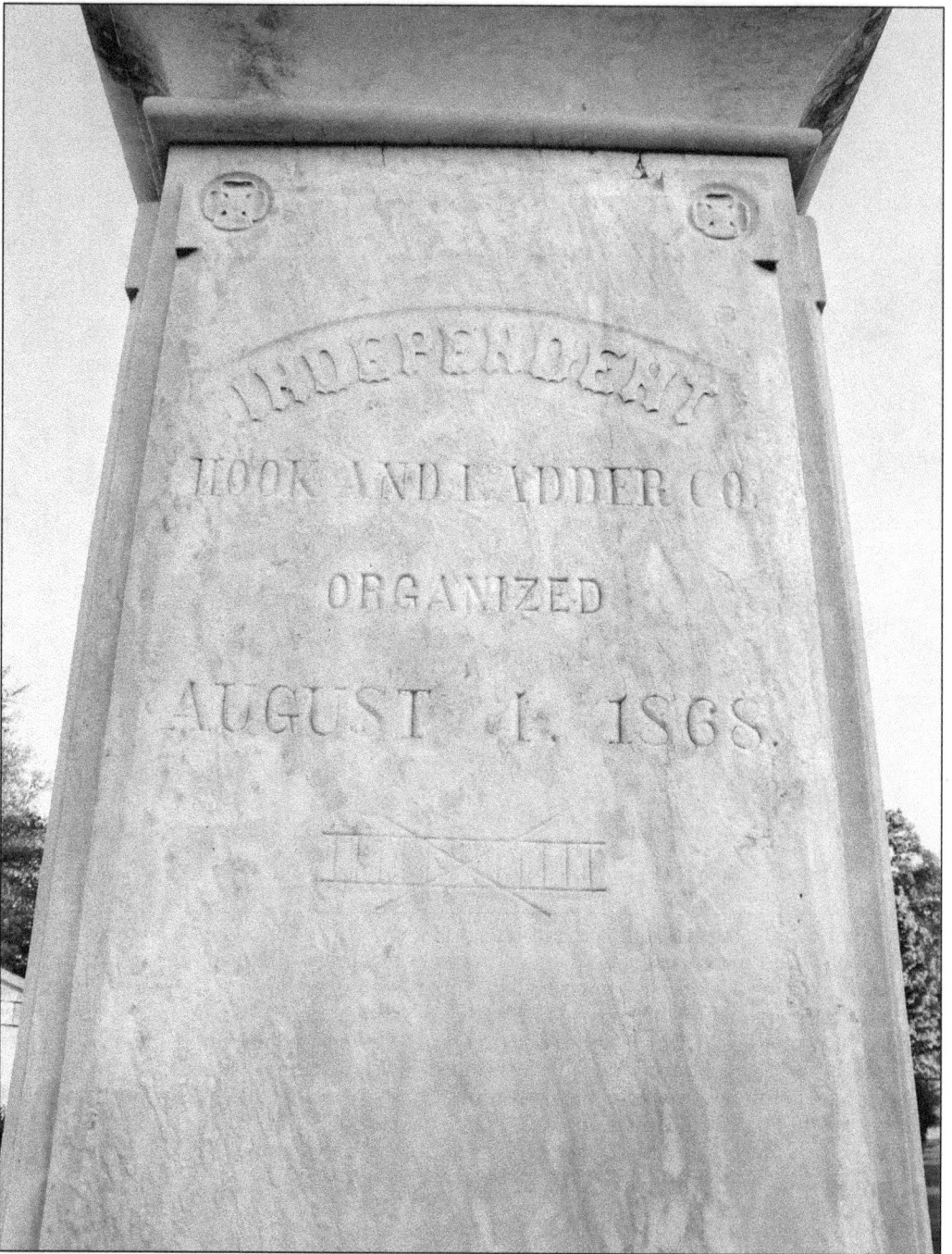

The final side of the monument honors the Independent Hook and Ladder Company, which was organized in 1868 to combat the challenge of taller structures in Charlotte. The horse-drawn apparatus of the hook and ladder consisted of a ladder that was hand cranked to rise along with several wooden ground ladders to reach victims. (Royall.)

Firemen pose with their horse-drawn steam fire engine, pictured here in the late 1890s. Later this apparatus was completely restored and is now proudly on display at the Charlotte-Mecklenburg Fire Museum. The firebox located in the rear of the steam fire engine was kept ready by the "stoker." This fireman would keep a small fire going that could easily be stoked at the call for duty by throwing a shovel full of coal into it. By the time the steamer rolled out of quarters, thick smoke would billow from the top of the steamer. (CFD Fire Museum.)

BIRD'S-EYE VIEW, LOOKING NORTH, CHARLOTTE, N. C.

A postcard from the late 1880s shows a bird's-eye view of Charlotte. The steeple-like brownstone structure on the right was city hall, and the firehouse was located in this building down the alley to the left. Located on the southwest corner of Tryon Street and Fifth Street, it stood from 1891 until 1924. (Charlotte-Mecklenburg Public Library.)

This actual photograph shows the imposing front of city hall, which housed the mayor's office, the police chief, and a firehouse, located down the left-side alley. (Charlotte-Mecklenburg Public Library.)

This panoramic photograph shows firemen standing in front of quarters *c.* 1911. The fireman in the center is the newly promoted chief of department, J. H. Wallace. Sixth from the left is Capt. G. W. Spittle, who would later die in a wreck involving his fire truck and a city trolley car. (Charlotte-Mecklenburg Public Library.)

Charlotte-Fire-Department.-Jan.16th 1916.

This 1916 photograph presents the same firehouse located on the side of city hall. By this time, the Charlotte Fire Department had progressed to nearly an all-motorized fleet of apparatus. Pictured to the far right, though, is the last of the horses in service. They would be retired in June of the next year. (Charlotte-Mecklenburg Public Library.)

Pictured here is chief of department Walter Orr in his horse-drawn chief's buggy. This beautiful buggy was shiny and solid black with gold leafing on the side that displayed "Chief, F.D. Charlotte." The buggy had leather seats, a lantern that hung in the rear, and a fire bell on the front to warn citizens of his approach while on a run. (Garris.)

This is a Charlotte chemical hose wagon from around the end of the 19th century. This apparatus carried 40 gallons of firefighting chemical along with a hose reel, extinguisher, and ground ladders. (CFD Fire Museum.)

Two of Charlotte's fire horses are pictured here in the 1890s. These horses were well cared for and became close companions to the firemen. Retiring the horses forever and moving to motorized apparatus was begrudgingly accepted by the loyal firemen but took several years to fully accomplish. These horses became so adept at their job, once the fire bell rang, they would proceed—often on their own—to their places in front of the apparatus. (Garris.)

Purchased in the early 1900s, one of Charlotte's first motorized fire apparatus was an American LaFrance engine. Pictured are three Charlotte chiefs of department. Bottom far left is a young Hendrix Palmer, seated to the right of the driver is the then-current chief of department J. Harvey Wallace, and standing at the rear far right is J. H. Wentz. After Chief J. H. Wallace was killed in the line of duty following a fire and subsequent dynamite explosion, Wentz would become chief of department. Hendrix Palmer would attain the same rank in 1927. (CFD Fire Museum.)

Taken in 1912, this early photograph depicts the Charlotte Fire Department during a tremendous time of change. The fire horses that had pulled the steam fire engines for so many years had recently begun to be removed from service as the firefighters began riding in motorized apparatus. Some items, however, had not changed, such as the leather fire helmets, lanterns, hose, and nozzles that still adorned the apparatus. It would not be until 1935 that the fire trucks would again see a drastic change for the better when enclosed cabs for the firemen to ride in were introduced. (CFD Fire Museum.)

This early 1900s photograph shows Capt. J. Harvey Wallace (right) with a young Hendrix Palmer (left) posing in their uniforms. The firefighters wore cap badges displaying the Maltese cross with their company number on it, while the officers wore cap badges displaying trumpets. (CFD Fire Museum.)

These Charlotte firemen pose in dress uniform for a department photograph while seated in front of quarters in 1916. A motorized fire truck with lanterns can be seen behind them. (Charlotte-Mecklenburg Public Library.)

Two

1920–1940

In the 1920s, the fire department and the city of Charlotte had grown not only outward, but also up. Skyscrapers dotted the inner city of over 46,000 residents, and the fire department was still evolving.

Likely the most influential person to serve the fire department, Hendrix Palmer moved to the chief of department position in 1927, thus beginning a legacy that would last until 1948. By 1930, there were 6 firehouses, 117 men, 7 engine companies, 4 ladder companies, a fire-alarm division, and a fire-prevention division. On July 14, 1930, Box No. 23 and Box No. 118 were struck for a fire in the Belk department store near Trade Street. The three-alarm blaze was finally brought under control with an estimated loss of $256,000. Fifteen firefighters were injured, one almost fatally. Fireman Westnedge plunged into the basement after the floor collapsed underneath his feet but was rescued by his brother firemen after they lowered a section of fire hose and pulled him to safety. Sadly, Westnedge would die in the line of duty four years later, in 1934, when the engine company he was riding in was involved in a wreck while responding to a fire at 117 South Cecil Street.

The Charlotte Firemen's Retirement System began in 1932. By 1935, there had been six line-of-duty deaths. The department purchased its first fully enclosed fire truck in 1935 after two of those deaths were a result of apparatus accidents. The department's 1937 annual report made the first mention of training, and by 1940, a drill tower and fire school were built and named after the chief who so passionately pursued its existence—Chief Hendrix Palmer.

In 1940, Charlotte's population had more than doubled since 1920 to over 100,000 people. Firefighters worked 84 hours a week, and Chief Palmer served as president of the International Association of Fire Chiefs. The city suffered its largest loss of life to fire on March 15, when a general alarm was struck for the Guthery Apartments at 508 North Tryon Street. Nine people died in a dramatic fire that saw multiple acts of bravery by the firemen. One firefighter was injured when a frantic woman jumped from her window and landed on him as he was attempting to raise a ladder to rescue her.

This panoramic photograph, taken on November 16, 1921, shows CFD apparatus and men on Morehead Street. Pictured at left is the horse-drawn steam engine, which by then was stored as a last-ditch reserve apparatus. Pictured on the apparatus fourth from left are "Fritz" and Captain

Palmer, and to their left was the fire alarm's Ford. The next car was an Essex chief's car carrying Chief M. Wallace. (Charlotte-Mecklenburg Public Library; photograph by Moon.)

Engine Company No. 2 and Ladder Company No. 2 are pictured in front of the original quarters, located at 1200 South Boulevard in the 1920s. Originally designed for horse-drawn apparatus, this was the first No. 2 built; it still stands today in the Dilworth area. Directly across the street is the second No. 2 built, which currently houses the arson task force. Engine and Ladder No. 2 now run out of a third firehouse, which is several blocks down South Boulevard from the first two buildings. (CFD Fire Museum.)

Pictured here is a 1935 seven-hundred-and-fifty-gallon Mack Engine Company enclosed cab. This was the first fully enclosed fire truck ordered by Charlotte. (CFD Fire Museum.)

Engine Company No. 7 is pictured here around 1920. This firehouse still stands in the North Davidson (No-Da) area of the city, which was once surrounded by mills and warehouses but is now considered the arts district of the city. Firehouse No. 7 originally contained a jail cell, which reportedly was used to temporarily house rowdy mill workers since this area was some distance from uptown and the police station. (CFD Fire Museum.)

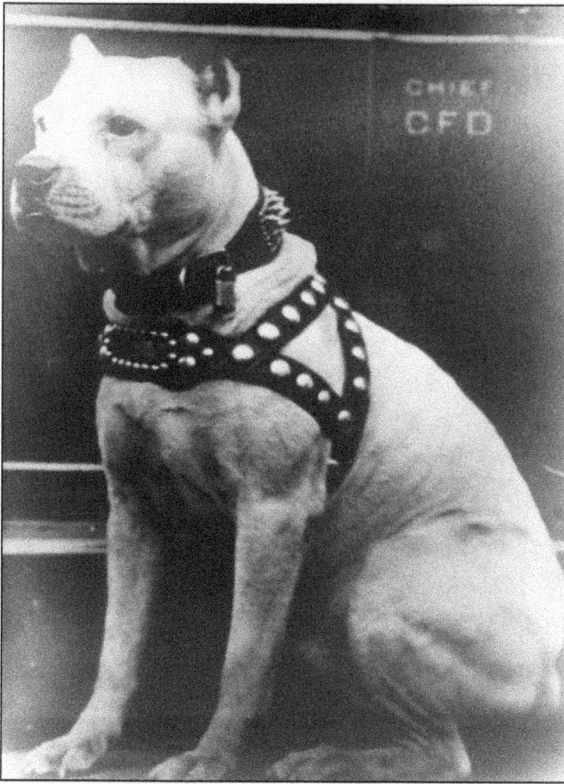

"Tony," the firemen's friend, sits beside the chief's car in the early 1920s. On Saturday night, September 26, 1922, Tony, a big, masterful bulldog and mascot of the central firehouse, was killed during the fighting of fire at the Carolina's Sporting Goods Company. Tony would run beside the fire truck and bark at anyone getting in the way of the firemen. On this night, the fire truck Tony was running beside suddenly veered into his path after being cut off by another fire truck, catching Tony off guard. He was run over and died at the scene. He is buried in front of the current Firehouse No. 4. (Garris.)

Capt. Hendrix Palmer is pictured here with Tony at his feet in early 1922. Tony's offspring "Fritz" would carry on as department mascot after Tony's death in 1922 but would suffer the same fate in 1926. (CFD Fire Museum.)

The original Firehouse No. 4 was located at 420 West Fifth Street in the 1930s. This former firehouse is the current home of the Charlotte-Mecklenburg Fire Museum. (CFD Fire Museum.)

This 1935 photograph shows Headquarters Firehouse No. 1. The chief drove an Oldsmobile sedan, while the assistant chief drove a Dodge sedan (on either end). The apparatus in the center are, from left to right, Engine Company No. 1, a Mack 750-gallon enclosed triple combination pumper; Engine Company No. 8, an American LaFrance 1,000-gallon triple combination pumper; the lighting truck; and Truck Company No. 1, an American LaFrance 75-foot tractor-drawn aerial truck. (CFD Fire Museum.)

Headquarters Firehouse No. 1 is pictured here in the mid-1920s. From left to right are the chief of department's staff car, the assistant chief's car, Engine Company No. 1, Engine Company No. 8, and Ladder Company No. 1. (CFD Fire Museum.)

Firehouse No. 3, located on Belmont Street, is seen here with its crew and apparatus in 1929. (CFD Fire Museum.)

This photograph shows Firehouse No. 3, located on Belmont Street, in the 1930s. Engine Company No. 3 was a 750-gallon American LaFrance triple combination pumper. (CFD Fire Museum.)

FIRST INCLOSED PUMPING ENGINE IN U.S.A.

OWNED BY

CHARLOTTE FIRE DEPT.

·1935·

This Charlotte Mack 750-gallon triple combination engine, pictured in 1935, was the first fully enclosed pumping fire engine built in the United States. It served as Engine Company No. 1 until its unfortunate demise from lack of foresight—it was sold for scrap. (CFD Fire Museum/Garris.)

Engine Company No. 5 is pictured in front of quarters located on Tuckaseegee Road. (CFD Fire Museum.)

Pictured at this fire department dinner, held upstairs at city hall, is Hendrix Palmer, seated next to the head of the table on the left at the table to the right-hand side of the photograph. (Charlotte-Mecklenburg Public Library.)

Pictured at above left is District Chief James "Jake" Brown, who served the department for 40 years, from 1937–1977. At above right is a 1930s photograph of a young fireman Brown. (Brown family.)

This 1935 photograph shows Ladder Company No. 3 and Engine Company No. 6 pictured in front of Firehouse No. 6, which is still located on Laurel Avenue. (Garris.)

Engine Company No. 3 is pictured in 1932 in front of Firehouse No. 3, located on Louise Avenue. (CFD Fire Museum.)

Ladder Company No. 3, an American LaFrance aerial, is pictured in front of Firehouse No. 6 in the mid-1930s. (CFD Fire Museum.)

The Charlotte Fire Department Chorus, under the guidance of director Milton Pannetti, is pictured above in 1940. The Charlotte Fire Department also had a fife and bugle corps to round out its musical display. They have both been replaced today by a pipe and drum band that continues to play events while representing the department. (Charlotte-Mecklenburg Public Library.)

This photograph shows Assistant Chief Griswold stepping into a 1948 Mack Engine Company truck parked in front of Firehouse No. 1. (CFD Fire Museum.)

Engine Company No. 7 is pictured in front of headquarters in the 1940s. This truck was reportedly a Dodge 500-gallon triple combination pumper. (CFD Fire Museum.)

A firefighter drives a team of horses pulling the steamer in a parade during the Mecklenburg Declaration Day celebration. (Charlotte-Mecklenburg Public Library.)

This photograph shows the flower-adorned grave site of a fallen fireman who was killed while operating at Box No. 33. (Charlotte-Mecklenburg Public Library.)

Three

1940–1960

In 1941, Morris Field Airport was dedicated with Mayor Fiorello La Guardia of New York City and Governor Broughton of North Carolina as principle speakers. The need for aircraft rescue and firefighting units was beginning. World War II began, and 63 Charlotte firefighters eventually served in the war. Fortunately for the Charlotte Fire Department, the 1940s were relatively slow and passed with little incident. In 1946, firemen were granted one day off a week and a 15 percent raise. Chief Palmer ended his tenure in 1948, and Chief Donald Charles succeeded him. Firehouse No. 8 opened, and No. 2 built new quarters after the existing firehouse, originally constructed for horse-drawn apparatus, became too small. By 1950, the city of Charlotte encompassed 31 square miles and had a population of 134,000. The fire department employed 192 and had 10 engines and 5 ladder companies that responded to 2,240 runs. A new 20-circuit master manual municipal class "A" fire alarm system was put into service, replacing the old class "B" 10-circuit systems that had been in service since 1925. Recruit training began, with the first class running in 1952. Recruits spent two weeks at the drill tower, went through the Red Cross First Aid program, and took a course in pump operations.

Firehouse No. 9 opened in 1954. A massive fire destroyed the Southern Railway Terminal. The initial alarm was turned in by Chief Charles, who saw the smoke from headquarters. Eventually the fire required a general-alarm response. In 1957, Chief Charles served as president of the IAFC, and Firehouse No. 10 opened its doors.

Chief Hendrix Palmer began his career as a volunteer in the early 1900s and quickly rose through the ranks of the Charlotte Fire Department, ultimately obtaining the rank of chief in 1927. He held this rank until his retirement in 1948. Along the way, he was instrumental in creating the first training site for the CFD, developed the first enclosed fire engine, and served as president of the International Association of Fire Chiefs in 1940. He has been without a doubt one of the most influential members of the department since its inception. (CFD Fire Museum.)

On March 20, 1956, two firefighters try the Self Contained Breathing Apparatus (SCBA) at the drill tower.

The grand opening of Firehouse No. 10 on Remount Road took place on May 16, 1957. Pictured is Capt. E. F. Dixson (Engine Company No. 13) showing the building and Engine No. 10's Seagrave 1,000-gallon triple combination pumper to neighborhood children (from left to right) Steve Cabanis, Al Robinson, and Mike Wilson.

This June 10, 1959, photograph shows the Seagrave apparatus of Engine Company No. 11 departing quarters on Twenty-eighth Street for a run. Also housed at No. 11 during this time was Engine Company No. 14.

Engine Companies No. 4 and No. 1 sit idling at the scene of an incident on May 6, 1959. The "bullet nose" Seagrave became synonymous with the 1950s and 1960s.

On November 29, 1952, this engine company departs the second Firehouse No. 2 on another run.

This 1950s photograph shows the American LaFrance truck of Ladder Company No. 1 with its aerial raised in uptown. If an elevated stream of water was used on these trucks, the firemen would stretch a section of hose to the tip of the ladder.

Steamer No. 3, "Old Sue," is being moved by the communications line crew on March 29, 1957.

An Engine No. 10 firefighter poses aboard Steamer No. 3, "Old Sue," which was in temporary storage at the Morris Field Airport firehouse in July 1956. (Charlotte-Mecklenburg Public Library.)

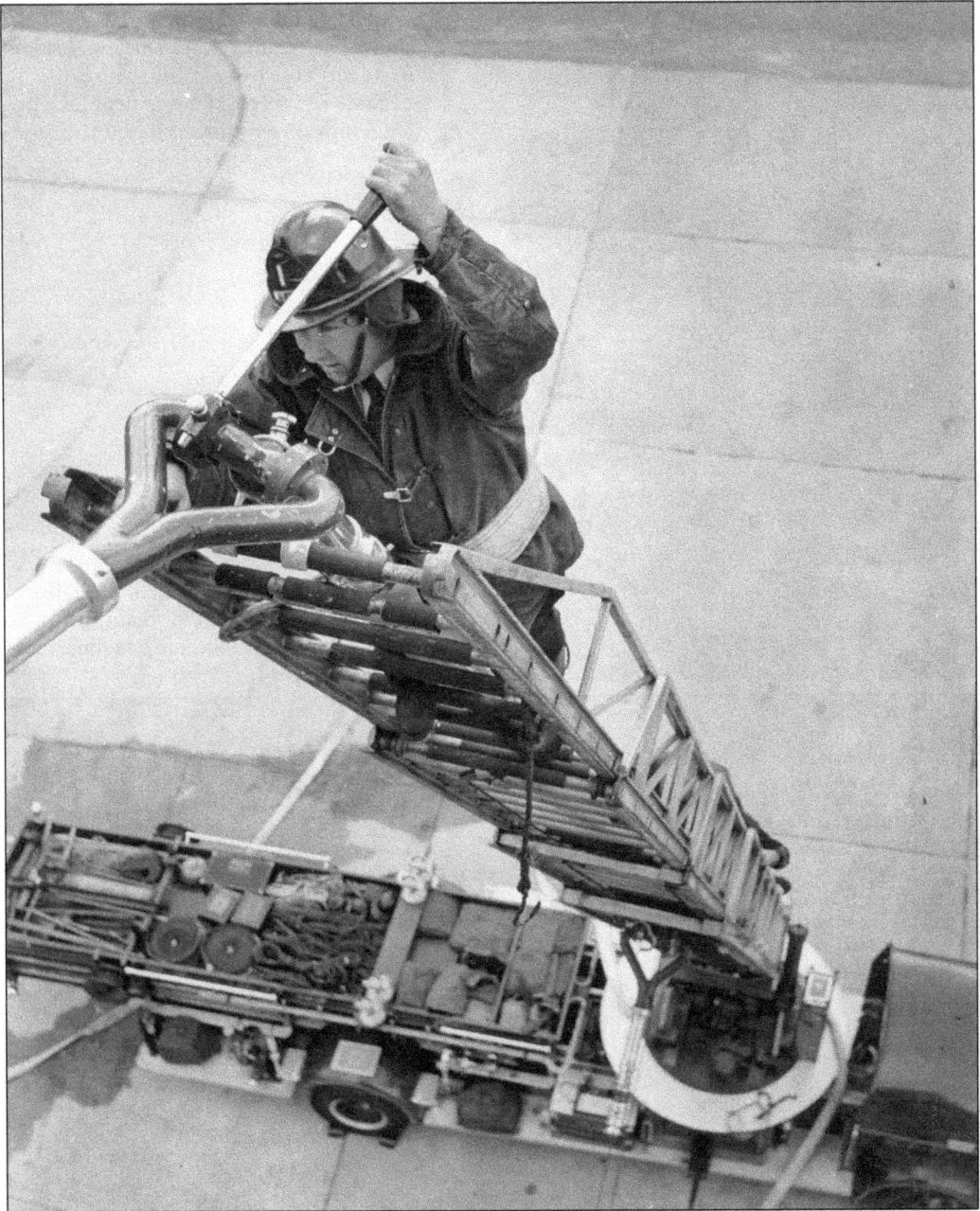

This 1957 photograph shows a firefighter from Ladder Company No. 1 directing a water stream from the tip of the aerial.

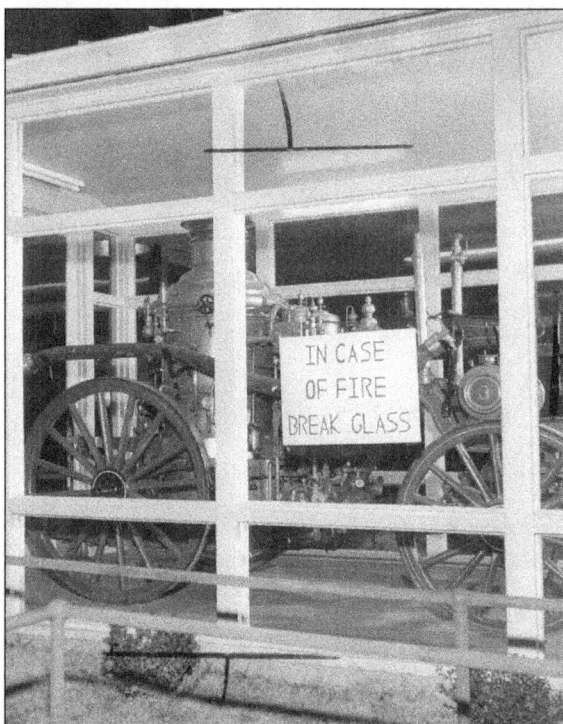

A photograph from January 31, 1958, shows the steamer "Old Sue" on display in its protective enclosure, which became the scene of a practical joke by local kids when they displayed this sign.

Although long since retired, firefighters clean and polish the old steam fire engine in the alley beside headquarters Firehouse No. 1.

Pictured is fireman Jim Jamison, who would eventually move up the ranks to serve as a district chief.

On August 25, 1959, an obviously exhausted firefighter, Don Black, in SCBA and an unidentified deputy chief discuss the fire at C. R. West Supermarket, located at 802 North Graham Street.

On June 24, 1954, Chief Charles saw smoke from his office and turned in the alarm for what would become a general alarm response to the Southern Railway Terminal fire. (UNC-Charlotte.)

On September 28–30, 1953, students and instructors attend a statewide apparatus pump operator school being held on the grounds of the Palmer Fire School. They are pictured here in front of the drill tower.

Standing in front of their American LaFrance fire truck in Firehouse No. 1 are the officer and firemen of Ladder Company No. 1.

During the 1950s, firemen practice on pompier ladders at the drill tower located on the grounds of the Palmer Fire School. No longer in use on current apparatus, this scaling ladder was used to climb the exterior of a building out of reach of ladder companies to get to victims. They are still used in recruit training to help recruits build confidence in themselves and their equipment. (CFD Fire Museum.)

This early model airport crash-fire-rescue apparatus was assigned to Morris Field fire station.

This photograph shows the American LaFrance of Engine Company No. 2 in front of quarters in 1949. (CFD Fire Museum.)

Pictured here is the Seagrave fire engine of Engine Company No. 6 after being struck by a City Coach bus in the area of Queens Road while responding to a box alarm.

Pictured here in 1956 is the CFD baseball team. (CFD Fire Museum.)

The fire department had very good baseball players in the 1940s and 1950s and often played all over the state. The love of the game continues today in the fire department, which also has a hockey team and a basketball team. Pictured above is a 1940s CFD baseball team. (CFD Fire Museum.)

This barbecue was hosted by the Don Hill Insurance Company at the Palmer Fire School in the early 1950s. Seated third from left is Chief Donald Charles. In addition to training, the Palmer Fire School was often used for gatherings such as dances, barbecues, and social events.

Pictured is the high-pressure unit operating at the Doggett Lumber Company fire. (UNC-Charlotte.)

This 1950s photograph shows the equipment used in communications, located in Firehouse No. 1. (CFD Fire Museum.)

Chief Palmer stands with two other firefighters on the front steps of the Palmer Fire School building. (CFD Fire Museum.)

Pictured in the 1950s are firefighters training in the drill tower with early-model SCBA equipment.

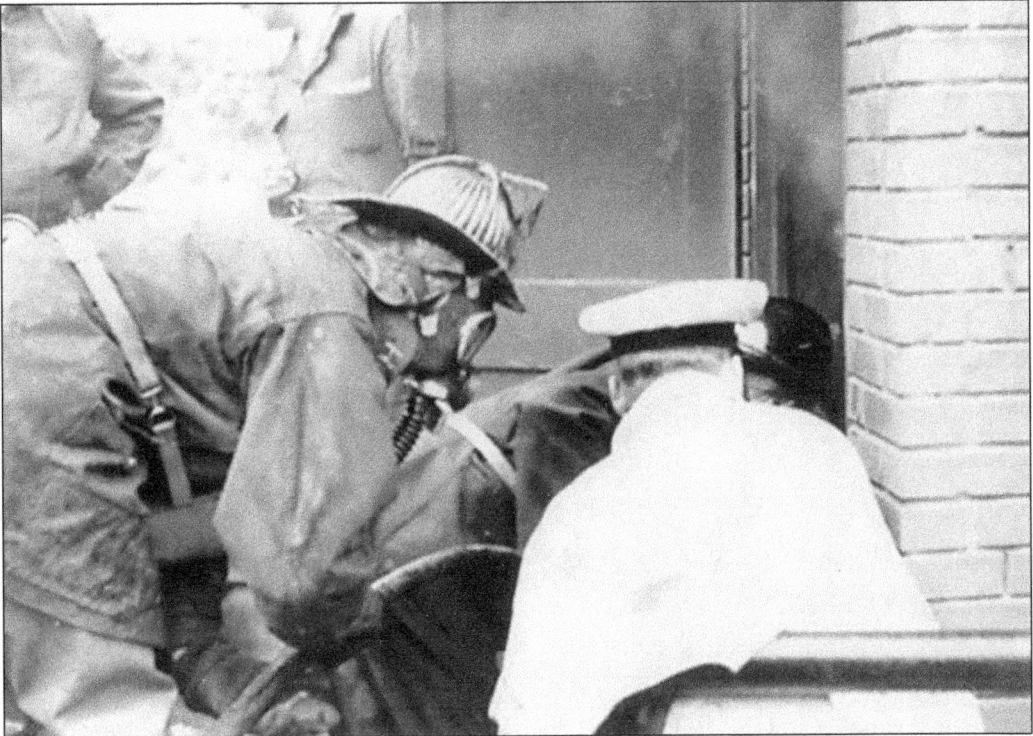

Platform No. 1 is pictured in front of the Palmer Fire School drill tower in the 1950s. (CFFA L660.)

Firefighters would often conduct high-expansion-foam training in the area between the Palmer Fire School building and the drill tower. (CFFA L660.)

Chief Griswold stands with members of city government in front of a 1948 Mack Engine Company truck. (CFD Fire Museum.)

Chief Palmer stands in front of No. 1 with an enclosed Mack engine and the light unit. (CFD Fire Museum.)

Firefighters operate two 2.5-inch hand lines while fighting a blaze in a commercial building in late 1959.

Shown sitting in front of Firehouse No. 8 are the Seagrave engines of Engine Company Nos. 8 and 24. (Holmes.)

This firefighter, perched precariously on top of cotton bales, searches for hot spots after a cotton warehouse goes up in flames in the 1950s. Note the sprinkler heads affixed to the trusses in the center, which, if working, no doubt did little to stop the fires progress.

The same firefighter from above sits on top of cotton bails while operating a hand line.

Pictured here is an exterior view of the same building with an engine company at left.

Here the high-pressure apparatus sits in the alley beside headquarters. Note the two deck guns to the rear of the truck.

This 1956 photograph shows the high-pressure unit flowing water at a fire along with other apparatus in the background.

An American LaFrance ladder truck is pictured here with its aerial being raised at the Palmer Fire School drill tower on March 20, 1956. (CFFA L660.)

Pictured in the center of this photograph is Chief Hendrix Palmer presenting an award to a firefighter. He is joined by his company in front of Firehouse No. 1. (CFD Fire Museum.)

This CFD fire alarm line truck was used to maintain the fire alarm pull boxes and associated equipment that was usually attached to power poles. (CFD Fire Museum.)

This is another view of the communications room at Firehouse No. 1. In 1961, the fire alarm system was maintained by a chief in charge of the division, five linesmen, nine dispatchers, and one record clerk. There were 846 fire pull boxes in service. (CFD Fire Museum.)

Capt. Glenn Shrum displays asbestos aircraft firefighting gear as others look on.

Four

1960–1980

The greatest impact to the Charlotte Fire Department occurred in 1960, when the city annexed 32 square miles and 10,000 homes. This required the installation of miles of new water mains and 300 alarm boxes along with the construction of six new firehouses over a six-year period. The year's largest fire occurred when an intense blaze burned the Southern Drug Company at 1400 East Morehead. In 1960, the city's population jumped to over 200,000. Walter Black became chief of the department, which established a computer system for fire department operations. A planning division was also formed.

The department struck another general alarm at the Doggett Lumber Company on October 19, 1964. The Charlotte Fire Department continued its example of professional service during the turbulent decade of the 1960s while simultaneously experiencing one of the largest growth periods in history. In 1966, a fire investigation division was organized, and in 1967, the CFD hired its first African American firefighter to serve the city since the days of the Neptune Company. By the end of 1969, the fire department was forced to reorganize into two divisions—operations and administration—and firefighters began working a three-platoon (shift), 56-hour week work schedule.

The period from 1970 to 1980 saw unique changes to the fire department—one being the hiring of the first chief of department that did not rise from within its own ranks. Chief John "Jack" Lee, hired by the City of Charlotte, oversaw the fire department through more annexation and, by 1979, the building of three additional firehouses in the midst of the city's financial shortfalls. The first firefighters to graduate from the Central Piedmont Community College Fire Protection Program emerged in 1971. On September 11, 1974, an Eastern Airlines flight crashed, killing 72 people on board. The single greatest change in emergency response to the Charlotte Fire Department occurred in 1978, when the firefighters became first responders and joined the Mecklenburg County Medical Emergency Response System. The end of the 1970s saw the removal of the Gamewell fire alarm pull boxes and the emergence of the 911 system, along with the growing abundance of telephones.

In 1960, firefighters work to extinguish the remains of this house fire.

One of the earliest known photographs of the CFD Honor Guard is this 1965 image, which shows the guard standing at attention during a fallen comrade's graveside service. Chief of department Walter Black joins them at far right. In 1965, the Honor Guard was comprised mainly of firemen who were ex-military, thus rifles were used along with the flags. Today's Honor Guard consists of 50 members who now carry axes, pike poles, and flags. They are joined by 20 members of the Charlotte Fire Department Pipes and Drums to make up a ceremonial unit that provides service at funerals, parades, and community events.

Airport Rescue Firefighting (ARFF) training takes place in February 1963 using high-expansion foam with an ARFF crash truck pictured at right.

A young child admires an asbestos ARFF suit on display in 1961.

Taken in 1961, this photograph shows the 1954 Seagrave apparatus of Engine Company No. 9 and firefighters in front of quarters. Engine Company No. 9 was in District 3 at this time and performed 363 runs in 1961.

The fire chief and deputy chief's four-door Buick sedans are parked in front of Firehouse No. 9 in 1962.

Engine No. 20 firefighters don their Scott SCBA while operating at this five-alarm fire on February 10, 1961, at the Southern Drug Company, located at 1400 East Morehead Street. Engine No. 20 was the second engine housed at Firehouse No. 1 along with Engine No. 1, Ladder No. 1, Deputy Car No. 2, Deputy Car No. 3, and Fire Chief Car No. 1.

Pictured is another view of the five-alarm fire at Southern Drug Company.

Engine Company No. 1 firefighters operate at the Southern Drug Company fire in 1961.

The 1959 Seagrave 100-foot aerial of Ladder Company No. 2 operates into the night at the scene of the Southern Drug Company fire in 1961.

Taken February 10, 1961, this photograph from the Southern Drug Company fire shows firefighters operating multiple hand lines while others attempt to ventilate the building.

A Ladder Company No. 5 firefighter attempts to bring light to the interior of the Southern Drug Company after the fire was brought under control. Behind him is a brass nozzle that was used to combat the fire.

The sun penetrates thick smoke at a fifth-alarm fire at the Southern Drug Company in 1961.

This photograph from July 1961 shows firefighters conducting night rescue training at the Palmer drill tower. A "victim" is wrapped in a tarp and tied between two pike poles that are attached to a roof ladder. The roof ladder is allowed to slide down an extension ladder while its decent is controlled by firemen using ropes from the ground.

On display during Fire Prevention Week in October 1961 is this 1952 American LaFrance of Ladder Company No. 1, at right, and other smaller versions.

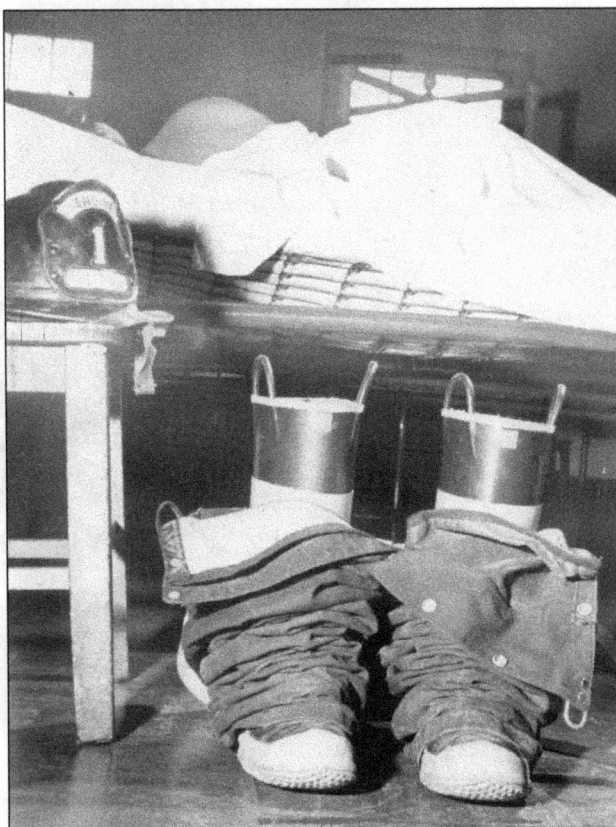

Taken September 23, 1961, this photograph shows an Engine Company No. 1 firefighter asleep "in the rack." Although firemen mostly used three-quarters-length hip boots during the day, they would often switch to "bunker gear" with suspenders at night.

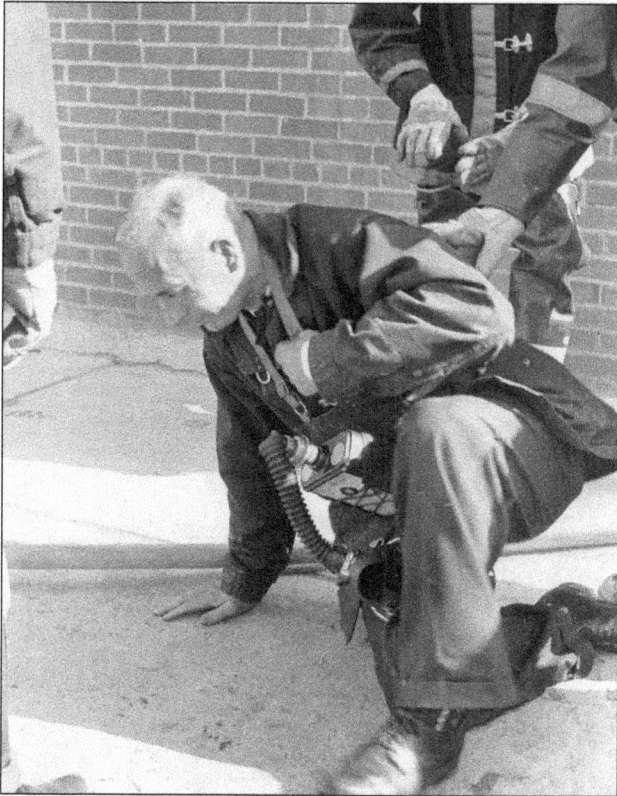

Capt. Frank Baker "takes a blow" after exiting the Southern Drug Company fire in February 1961.

This January 17, 1961, photograph shows Derita Volunteer firefighters and CFD firefighters aggressively advancing on a well-involved building fire. Take note of the painted logo above that exclaims, "Buy Insurance."

This is a photograph of the Doggett Lumber Company fire, a general-alarm fire fought in 1964.

This is the Doggett Lumber Company fire in 1964.

Engine No. 20 firefighters operate multiple hand lines at this building fire on August 16, 1963.

Here firemen conduct rope-rescue training at the Palmer Fire School drill tower.

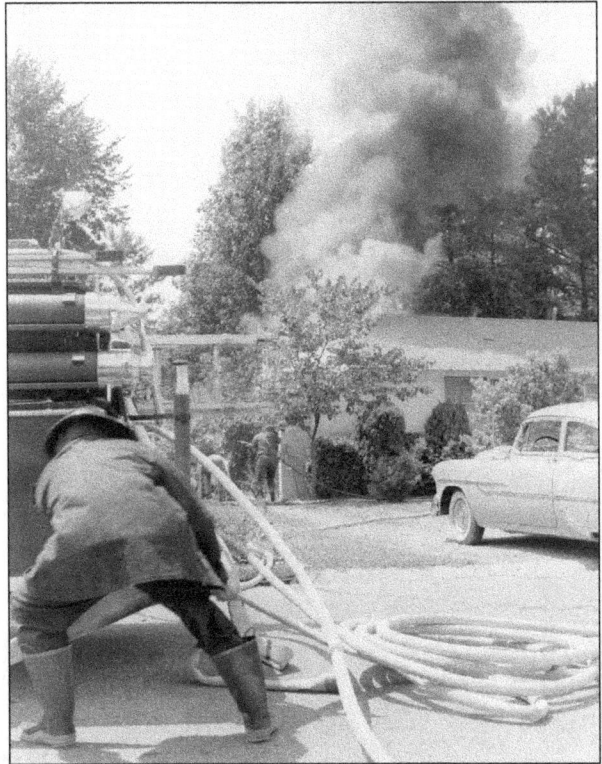

Taken on June 28, 1960, this photograph shows firefighters hooking up and stretching lines at a house fire.

Ladder Company No. 1 sets up their 100-foot aerial on a building adjacent to the Woolworth and Company building in uptown during a box alarm on February 9, 1961.

Newly appointed chief of department Walter Black (left) and outgoing chief Donald Charles (right) discuss the future of the fire department on July 24, 1963.

District chief W. O. Dowdy gives orders using his bullhorn, the modern-day trumpet, during a training exercise in February 1963.

This photograph, taken on April 29, 1964, shows firefighters perched precariously on top of a wall while operating two hand lines in an attempt to get water on the fire.

Running out of Firehouse No. 4, Engine Company No. 21 firefighters use high-expansion foam to extinguish the remains of this deep-seated fire on April 29, 1964.

The firefighter can almost be heard asking, "Hey buddy, got a match?" Here firefighters have a cigarette, or "take a blow," after hard work at an October 2, 1965, fire.

The Charlotte Fire Department Honor Guard stands at attention as Chief Walter Black assists the Mecklenburg County Volunteer Fireman's Association president in placing a wreath at the fireman's monument in Elmwood Cemetery.

These firefighters are silhouetted against the sky at this fire in May 16, 1967. As firefighters on the ground operate a deluge set, another firefighter operates the nozzle on the aerial.

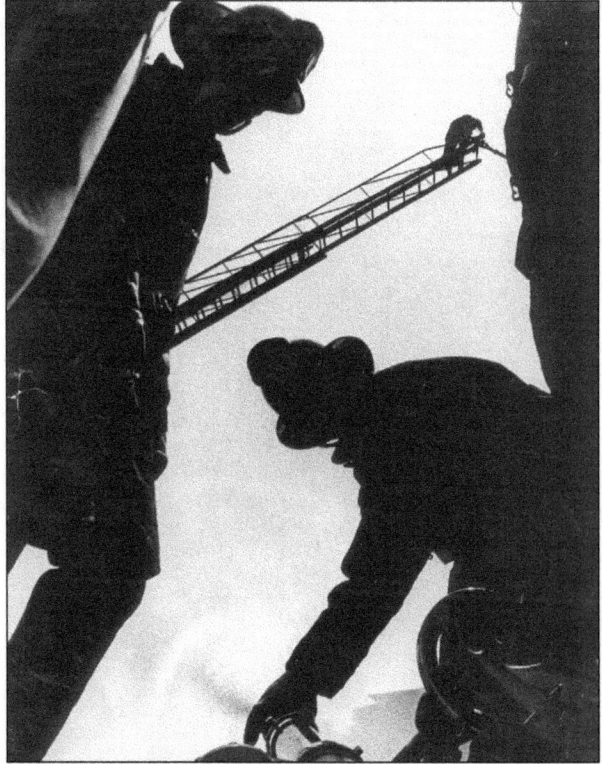

On March 4, 1965, three firefighters open up the roof to expose hidden fire at a house. Although today's firefighters usually use saws to accomplish such a task, the pick head ax is still carried and is often relied on because it always cranks!

Taken on August 6, 1966, this photograph shows firefighters as they operate hand lines into the building using ground ladders.

On May 16, 1967, firefighters operate multiple master streams at the Multiply Corporation fire at 4709 Rozzelles Ferry Road.

Two firefighters operate the master streams on the high-pressure unit at a November 2, 1965, fire as ice forms from the cold.

This 1962 Seagrave sits idle as the firefighters of Engine Company No. 2 operate at a car wreck on a rainy day, November 10, 1965.

Pictured in the late 1960s, this fire at the Penny Profit store was well advanced on arrival of the first engine company. Here the company stretches an attack line off their bullet-nosed Seagrave fire truck in order to knock the fire down. (CFD Arson.)

This photograph, taken on November 11, 1967, shows firefighters operating on a building fire at 900 Marsh Road that has extended to vehicles. (CFD Arson.)

Firefighters douse the remaining flames at a fire that broke out in the Piedmont Supermarket on Oaklawn Avenue on August 12, 1968. (CFD Arson.)

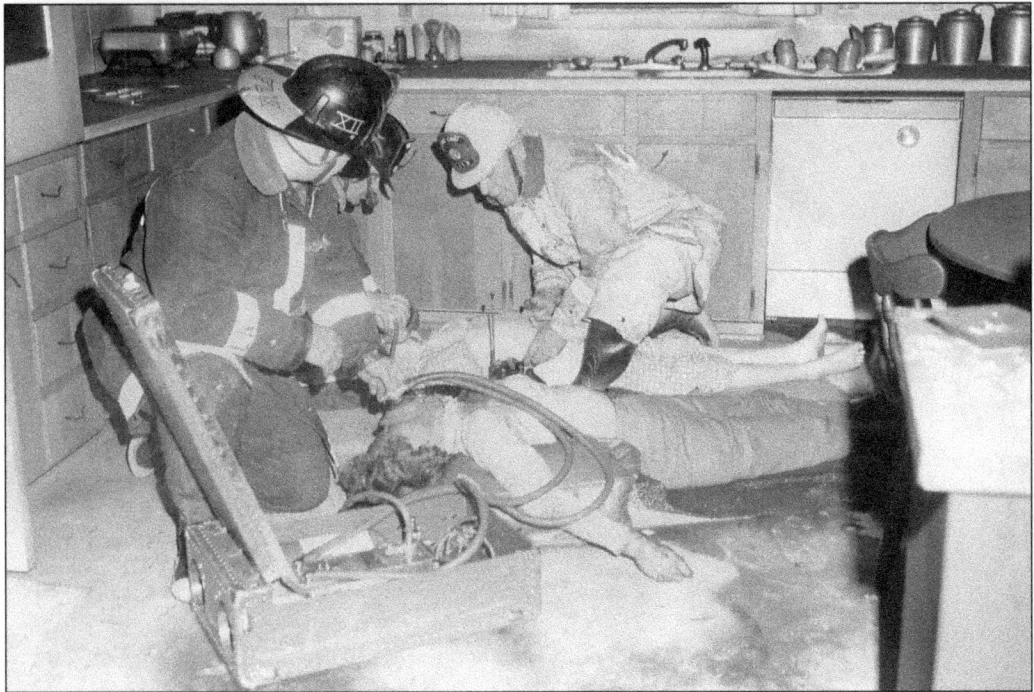

These firefighters work to get oxygen on two victims as a district chief attempts chest compressions. This arson-related fire was intended to cover up a murder-suicide at 3901 Sulkirk Road on February 16, 1968. (CFD Arson.)

A firefighter searches for hot spots after a two-alarm fire rips through the A&P building on Hovis Road. (CFD Arson.)

The 1959 Seagrave truck of Ladder Company No. 2, passing a parked Charlotte police car, displays a fire prevention sign during Fire Prevention Week on October 4, 1965.

Pictured here on January 10, 1966, an Engine Company No. 2 firefighter, Al Caudle, hoses down an exhausted Deton Oliver after fighting a fire at the Dixie Waste Mills at 701 West Palmer Street.

At a fire in the 1970s, Capt. Oscar White assists firefighter Barry Rhyne in hooking up hose lines, while Fred Lowe operates Ladder Company No. 2's aerial. Firefighter Glenn Swann is pictured at far right.

Pictured are firefighters Bill Snipes and Smokey Dorsey in 1972. (CFD Fire Museum.)

Firefighters on the aerial high above Ladder Company No. 2 operate at this building fire on January 28, 1972.

Pictured here on August 11, 1971, are, from left to right, firefighters John Fountain of Engine Company No. 14 and Ken Brown on the nozzle.

Taken on February 17, 1978, this photograph shows EMS runs beginning for the CFD. Here Capt. Ben Whitmore has his blood pressure checked by Capt. R. H. Widenhouse as firefighters Mike Stegall and G. R. Parsely observe.

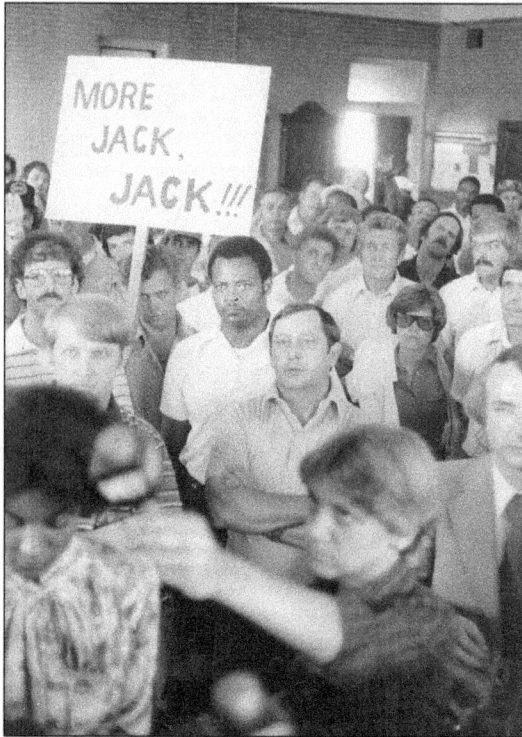

On June 8, 1979, over 100 firefighters marched on city hall to demand a pay raise. The sign reads, "More Jack, Jack," in reference to then–chief of department John "Jack" Lee.

CFD firefighters assist Oakhurst Volunteers with a brush fire off Plaza Road extension on February 28, 1974.

A firefighter dons his SCBA mask and prepares to enter the building at an apartment fire on December 23, 1976.

On July 6, 1979, less than a year into officially becoming EMS providers, two Charlotte firefighters provide EMS care to an elderly woman who was injured in a fall.

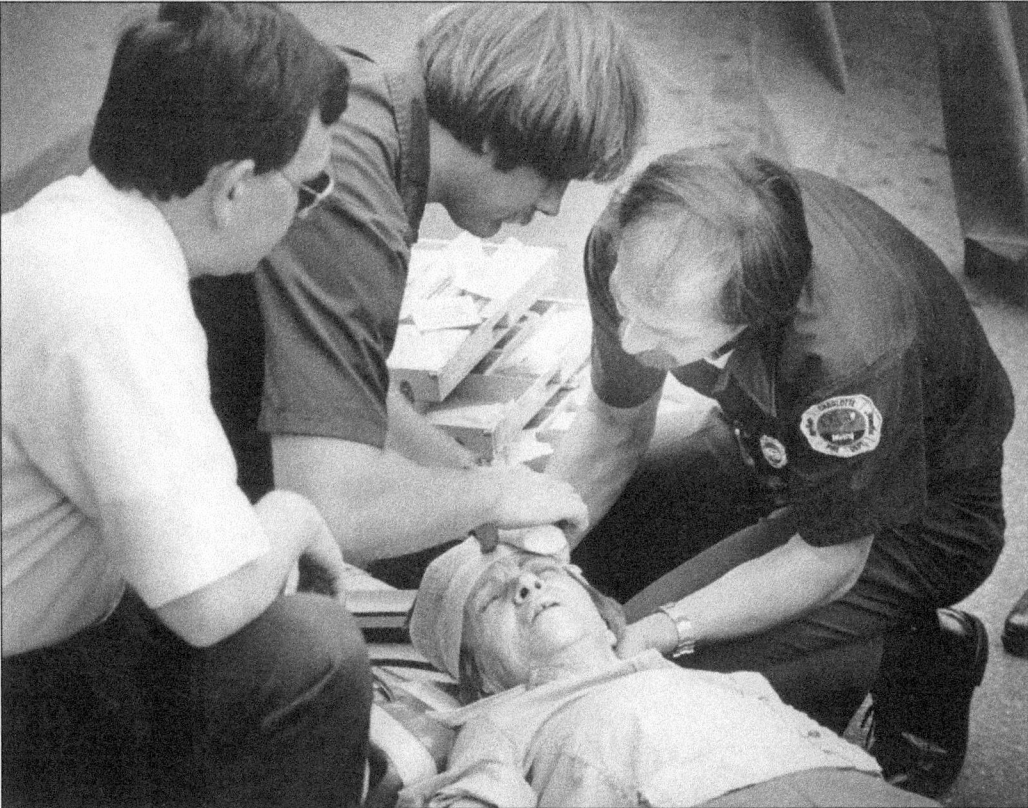

A fire department line crew checks a pull box to ensure it is working properly in 1976. All pull boxes were removed from service a few years later when the 911 system became effective. (CFD Fire Museum.)

The crew of Engine Company No. 6 poses in front of their 1500 Seagrave with their Dalmatian mascot.

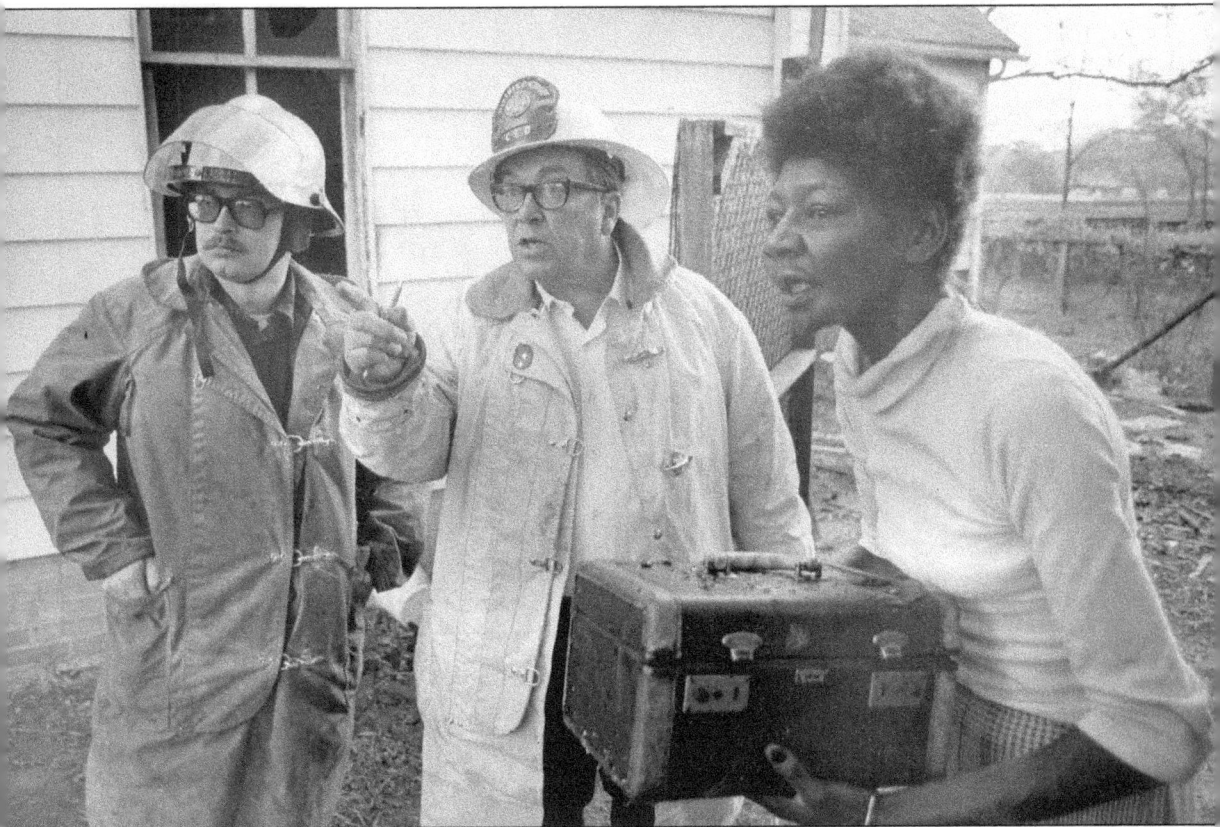

This April 18, 1979, photograph shows a captain and district chief helping a citizen after a fire destroyed her home.

On February 18, 1979, snow blankets the ground as firefighters operate at a fire in the Space Shuttle nightclub. Several generations of fire apparatus are pictured.

Firefighter Danny Phifer and Capt. Marion Baxley of Engine Company No. 3 work to extinguish a car fire on October 23, 1974.

Taken at a fire on July 24, 1977, firefighter Ken Brown of Squad Company No. 1 consoles a young girl after she has burned her hand.

Taken *c.* 1979, firefighters from Engine Company No. 1 take a much-needed rest after a commercial building fire.

Five

1980–1990

The 1980s could be referred to as "the decade of fire" as the Charlotte Fire Department responded to many multiple-alarm fires. By this time, the city covered an area of 137 square miles and the fire department was composed of 700 employees and 23 firehouses. Charlotte "smoke eaters" were wearing rubber boots, insulated overcoats, orange asbestos gloves, and sometimes SCBA bottles on their backs.

For the first time ever in the late 1970s, CFD trucks were painted yellow but fortunately did not remain so for long. In the fall of 1980, the department employed its first female firefighter. Although many large fires were fought during this decade, it will still most likely be remembered for the inception of the Hazardous Materials Response. After a number of fires in Charlotte involving hazardous materials, and particularly the Baxter-Harris fire in 1982, the fire department was tasked with finding a way to improve the department's response to hazardous materials. After the city compiled a list of recommendations regarding the its response to such incidents, Battalion Chief L. L. Fincher (future chief of department) was appointed Hazardous Materials Coordinator and given the responsibility of implementing the recommendations.

Chief Lee retired in 1982, and Richard "Blackie" Blackwelder became chief of department. His first action as chief was to rewrite the age-old city fire codes. The largest, single monetary loss fire to date occurred April 23, 1985, when a three-alarm fire destroyed the Royster Fertilizer Company, with estimated losses around $7 million. In 1988, the Charlotte Fire Department instituted a computer-aided dispatching system. The decade ended with the city bracing for the arrival of Hurricane Hugo. A category five hurricane at landfall, Hugo swept through Charlotte at the perfect time for emergency responders . . . in the middle of the night when everyone was off the road and asleep in bed!

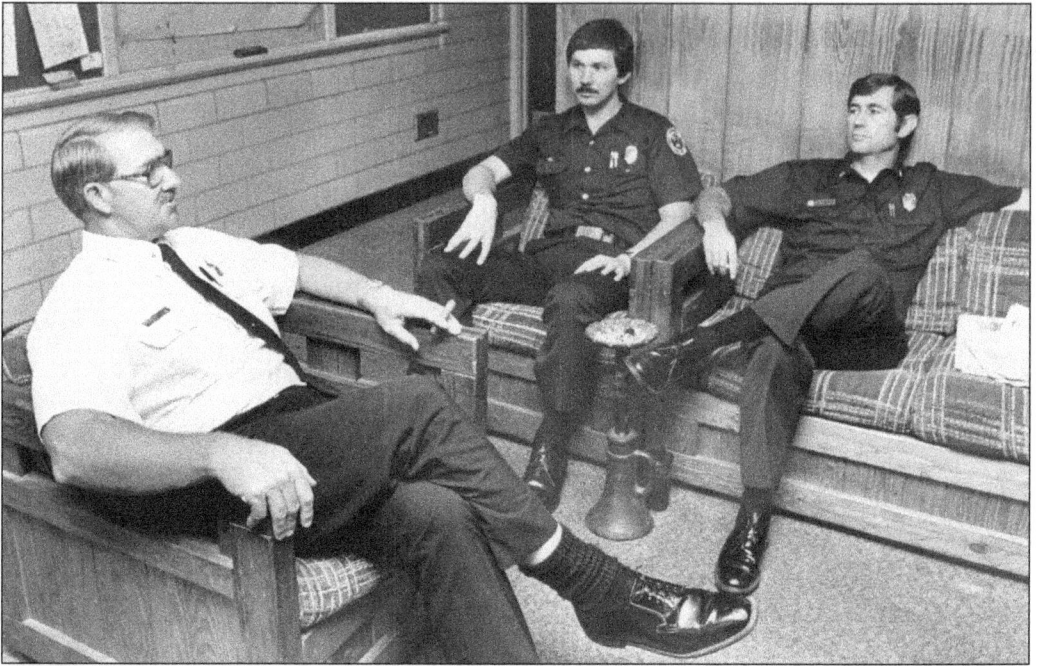

Much has changed in the fire department since this photograph was taken inside Firehouse No. 10 in 1981. Pictured are, from left to right, Chief Carson Watts, firefighter Roger Tench, and Capt. Charlie Tomlinson. Today tobacco use is not permitted inside the firehouse or inside any city-owned vehicles.

A firefighter enjoys the cool breeze while resting after this fire on June 10, 1989.

On May 4, 1988, firefighter Mike Wilson of Ladder Company No. 2 holds Brandon Snyder after his family was involved in a motor vehicle accident near Billy Graham Parkway and West Boulevard.

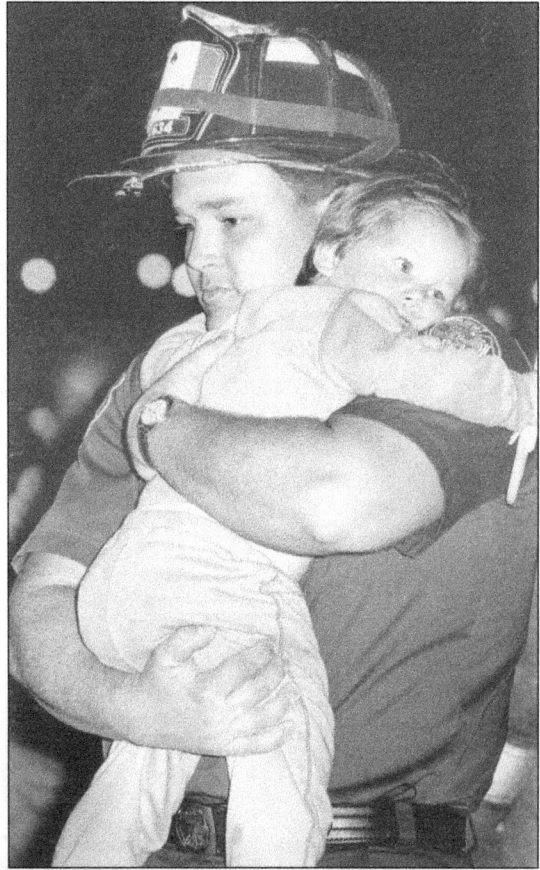

Taken on January 25, 1983, this photograph shows firefighters and paramedics working together to quickly move a victim after he has suffered cardiac arrest. Note the chest compression machine (thumper) attached to his chest.

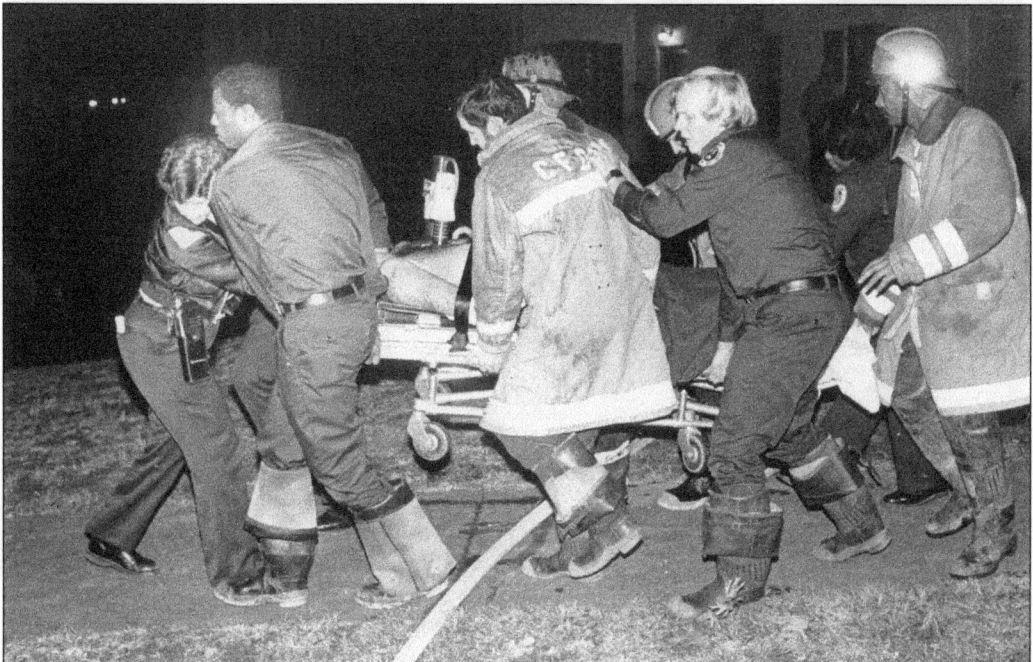

Arson investigators enter a house after a fire that resulted in a fatality.

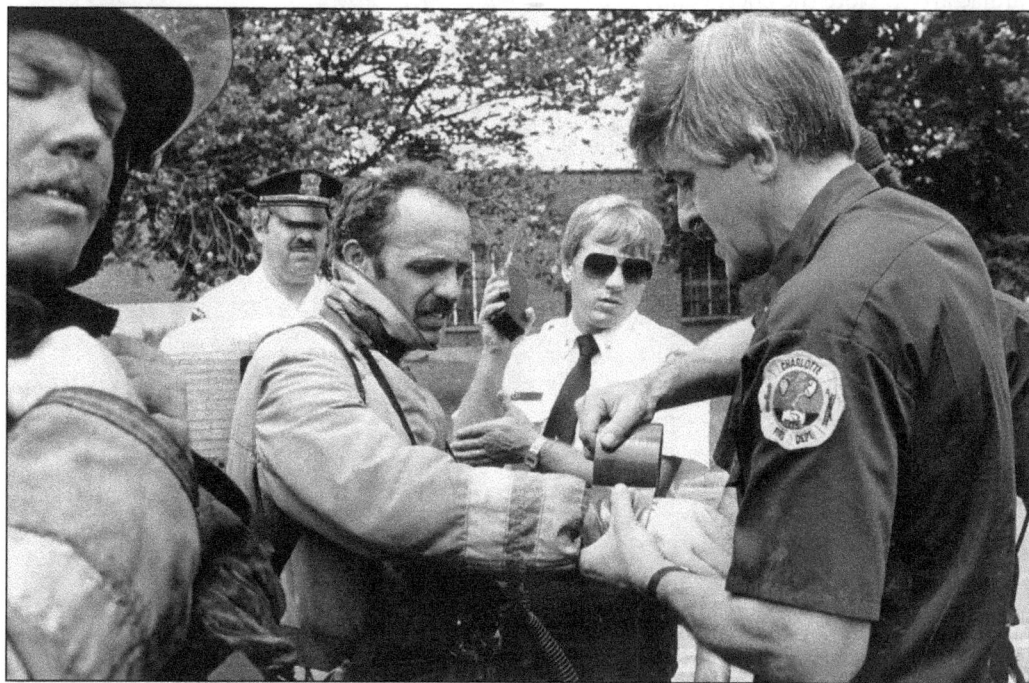

A firefighter has his protective clothing taped during a hazardous materials incident.

At a fire on November 11, 1981, Capt. Paul Sprinkle, lying beside Platform No. 1, was injured after a hose coupling disconnected and struck him.

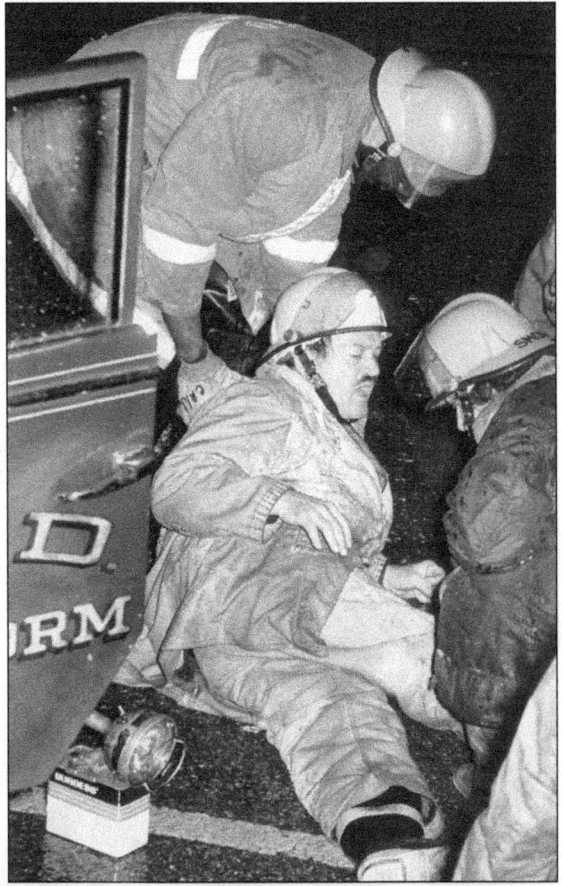

Arson investigators work at the scene of a fatal house fire on November 3, 1981, at 2704 Old Steele Creek Road.

This photograph, taken on January 20, 1986, shows firefighters operating at a commercial building fire at 418 West Fifth Street.

On August 1990, firefighters Ken Brady and Dwayne Sloan are checked over by paramedic Van Goodman after a working fire at 2601 Belvedere.

Firefighters check for fire extension at the Arboretum Apartments in 1991.

Ladder Company No. 4 operates at Wholesale Retreading on 1101 South Clarkson Street on May 8, 1984.

On March 1, 1982, firefighters operate a hand line at this fire at 630 South Graham Street, which consumed half a city block and a condominium under construction.

On April 29, 1987, this Snorkel Ladder Company operates at a fire at Providence Place Apartments, located at 1333 Queens Road.

On Christmas Day 1983, the temperature hovered around 12 degrees as many fires were fought throughout the city. The Furniture Mart on East Trade Street was one of the largest fires that day. After the fire was extinguished, the firefighters of one ladder company discovered the aerial had frozen in the up position; it was eventually driven back to quarters to thaw out and be lowered.

Ladder Company No. 18 and Engine Company No. 2 work the back side of the Furniture Mart building fire on Christmas Day 1983.

This photograph was taken at a fire that destroyed the Bryant Heating and Air building on December 11, 1987. It was reported that one ladder company tipped over at this fire, but no firefighters were injured. (CFD Arson.)

This photograph was taken on November 12, 1981, at a fire in the Pritchard Memorial Baptist Church. (CFD Arson.)

In this photograph, taken at a fire on March 10, 1982, firefighters work to place a deck gun into operation.

A fire destroyed the Providence TruValue Hardware Store on December 9, 1982. The photograph shows multiple hand lines and ladder companies operating to extinguish the fire. (CFD Arson.)

This photograph shows a two-alarm fire in the Dilworth Theater, which was fought on January 10, 1984.

This April 12, 1987, photograph shows a working fire at the East Pointe Apartments. (CFD Arson.)

A spectacular three-alarm fire destroyed Crockett Baseball Park, home to the Charlotte O's, on March 16, 1985. The size of the flames can be compared to the telephone poles seen standing in the center of the photograph. Property loss to the old wooden ballpark was estimated at $1 million. (CFD Arson.)

This aerial photograph shows a two-alarm fire that destroyed the H and S Lumber Company on October 21, 1992.

This photograph, taken on May 15, 1986, shows firefighters placing their deck gun into operation during a defensive fire attack at the *Weekly Newspaper* building.

Ladder Company No. 4 prepares to send water through its aerial to this fire, which consumed the First ARP Church on November 14, 1989. (CFD Arson.)

On May 28, 1987, this fire in the Grice Showcase building challenged firefighters.

On February 21, 1988, a massive fire at 3400 Rozzelles Ferry Road burns, and a ladder company prepares to raise its aerial and place its heavy master stream into operation.

On August 25, 1984, a car carrying a store robbery suspect collided with a gasoline tanker at 5000 East Independence Boulevard, resulting in a three-alarm fire. (CFD Arson.)

On May 27, 1981, District Chief R. J. Ellison assumes command at a fire at Cumulus Fibers on Tarheel Road.

Taken August 4, 1983, this photograph shows the Mack TeleSquirt apparatus of Engine Company No. 13 operating at a fire in the Western Sizzlin restaurant near Freedom Drive and Tuckaseegee Road.

Pictured are hazardous materials firefighters from Engine Company No. 13, operating in protective clothing on top of a tank as they work to contain an acid spill at Worth Chemicals on January 28, 1990.

Firefighters ascend the aerial of Ladder Company No. 4 at this building fire.

A firefighter directs a stream as this ladder company operates at a church fire in 1981.

On November 24, 1983, firefighter Benny Clark operates the nozzle while keeping an acetylene tank cool at the Westinghouse Plant located at Eastway Drive and Sugar Creek Road.

This photograph from 1985 shows firefighter Larry Starr of Engine Company No. 5 advancing a line on a car fire.

In 1980, firefighters discard a burnt mattress from an apartment at Piedmont Courts after a fire was started when someone fell asleep while smoking in bed.

Taken on November 11, 1981, this photograph shows Ladder Company No. 15 operating at the J. King Harrison fire on Brevard Street.

A fire on December 25, 1983, consumed the Furniture Mart on Trade Street.

Taken on August 9, 1983, this photograph shows firefighters having to contend with downed power lines and attempting to save the home on the right after the house next door burned to the ground.

This 1988 photograph shows Capt. Oscar White of Engine Company No. 1 as he advances the line to his nozzle man at a house fire on Mint Street and Carson Boulevard.

FURTHER INFORMATION

Located in the historic Fourth Ward district of Charlotte, the Charlotte-Mecklenburg Fire Museum and Education Center is housed in the original Firehouse No. 4, which was built in 1925. It closed in 1972. The building lay dormant for several years until it was reoccupied by firefighters in 1999, when the Charlotte Fire Department opened the Charlotte-Mecklenburg Fire Museum. This was pursued in order to provide a home for those who sought to preserve the history and restore the existing artifacts to their original glory. The museum is maintained by a dedicated group of both active and retired Charlotte firefighters and their families. As high-rise buildings continue to be constructed around this old brick building, it serves as a reminder of days gone by. In recent years, through the tireless efforts of many Charlotte firefighters, several antique fire trucks and equipment have been restored to their former glorious state and are on display inside the fire museum. They occasionally still rumble down the streets of Charlotte during events such as parades. Pictured along with the building that houses them are several of Charlotte's restored apparatus that were featured in this book and that served the citizens of Charlotte for many years.

This photograph shows the original Firehouse No. 4, which is now home to the Charlotte-Mecklenburg Fire Museum and Education Center.

This hand-drawn hose reel is similar to what was used by the volunteer firemen in Charlotte before 1887.

The 1901 horse-drawn American first class steam fire engine that served Charlotte until around 1916 is now affectionately known as "No. 3 Old Sue." Completely restored by Charlotte firefighters, this steamer is now on display at the Charlotte-Mecklenburg Fire Museum. "Old Sue" has stood the test of time and recently pumped a stream of water when a group of firefighters relit its firebox.

This *c.* 1929 American LaFrance 1,000-gallon triple combination pumper was ordered in 1928 and served the City of Charlotte for well over 20 years. Completely restored by firefighter Jeff Dixon, it is now on display at the Charlotte-Mecklenburg Fire Museum.

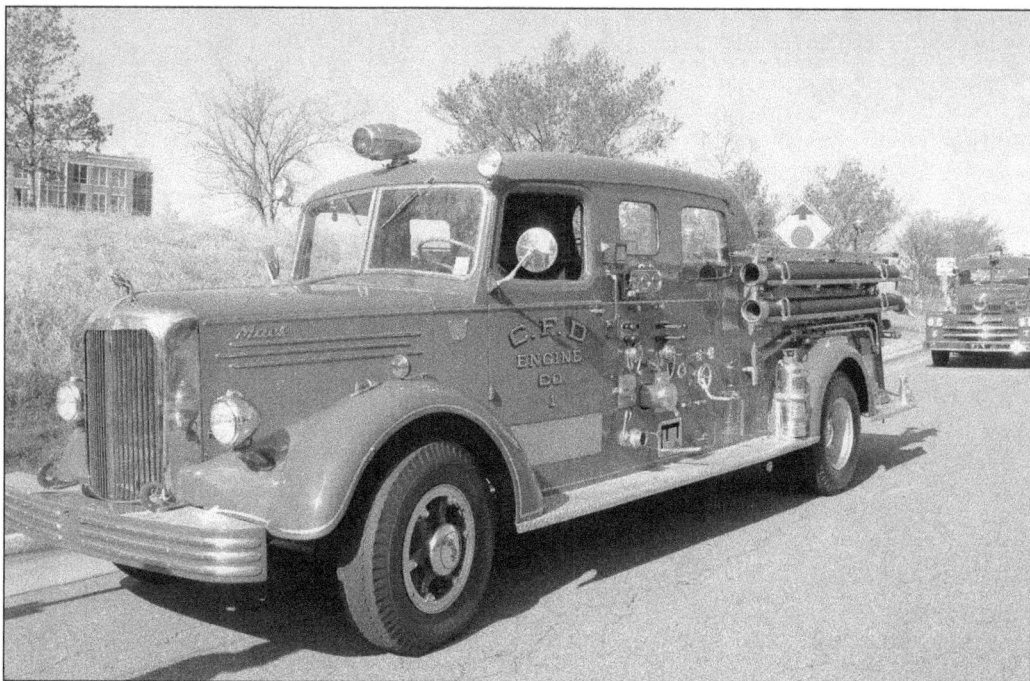

This 1948 Mack 750-gallon triple combination pumper was retired from service as CFD Engine Company No. 6. Completely restored, it is on display at the Charlotte-Mecklenburg Fire Museum.

This photograph shows a 1958 Seagrave 1,000-gallon triple combination pumper, which was retired from service as Engine Company No. 23. It quartered with Engine Company No. 11 in a double-engine house at Firehouse No. 11. It has been restored and, although it is driven in parades and often on display, remains in storage because of limited space at the fire museum.

This 1971 Seagrave 1,000-gallon pumper retired from service as Engine Company No. 16. Although it has been restored and is driven in parades and often on display, it remains in storage because of limited space at the fire museum.

Visit us at
arcadiapublishing.com

www.ingramcontent.com/pod-product-compliance
Lightning Source LLC
Chambersburg PA
CBHW080631110426
42813CB00006B/1656